Caught in Between

Understanding why we are too connected to unplug, and too old-school to live online.

Dedicated to my brothers and cousins for pushing me to learn everything about MTV, Nintendo, and programming way before I had to.

Contents:

Preface: The Quantum Leap   iv
Introduction: Too Much Information   vii

1. Fearless and fearful digital behavior.   10
2. The Remote-Control phenomenon   22
3. The Frankenstein Phenomenon   27
4. It takes a village   31
5. Keep it to yourself - privacy   35
6. Bonus points and reset buttons   45
7. More than words   52
8. TTYL, BFFs, and Tinder   57
9. Guarding your soul in a world of likes.   63

Preface: The Quantum Leap

Do you know the names to all the trending shows and artists nowadays? How many apps do you have on your phone and tablet? Have you ever caught yourself using your cell phone or tablet while you are watching TV?

If you were born in the 70s or 80s chances are the TV was your nanny; and you still remember the songs, the characters, shows, and stories you watched over and over again. Maybe you played videogames in a video arcade, or you owned an Atari console, not to mention floppy disks. Maybe you know the thrill of holding a joystick, and can quote anything from *Back to The Future*, *Star Wars*, and many, if not all *Disney* movies.

If you were born during this time frame, you are also one of the privileged human beings who was able to feel comfortable with technology, but just enough to miss playing in the streets or having to face the thrill making a phone call with no caller ID.

As time went by, there was a "Quantum Leap" in technology if you know what I mean. We went from phones that resembled bricks and floppy disks; to a phone that has more memory in a tiny card than a whole pile of those disks. We went from film rolls and disposable cameras, to having our whole lives recorded online in HD. If you have lived through that, we have been riding the same wave. But if you lived that, you may now be too busy to keep up because... There were diapers, meals, packing lunch, going to work, getting clothes ready, planning vacations, meetings, health crisis, and plenty of other things to do. All of which, developed faster than the Ice Bucket Challenge turned into the Laundry Pod Challenge.

I, however, just grew up and went from playing with technology and enjoying pop culture, to understanding it for a living. I rode the wave from the beginning. I joined Facebook when it was basically an online yearbook, and I have witnessed the good, bad, and ugly that has come with it. I have worked with influencers and public figures. I started working with digital communities when they were still just for famous people or famous topics; not knowing that it would be something as necessary for parents as helping their children with homework.

I want to give you a competitive advantage and insight into a world that your children will have to face and you may not understand. Because we did have those years when we were free to play, to fail, and to build relationships face to face. And yet, we know the thrill of technology, the adrenaline of getting that extra life in a video game, or making a line to be the first to watch the last movie of the trilogy. We understand that movies and TV have become as influential to us, as the classics and ancient wisdom became to humanity.

Now, we are educating children that will rarely understand the difference between private and public. They may never experience a time in their life when they were "unplugged" or "disconnected" and just enjoy being. They may not be able to live a "simple life" as a means to survive. They need us to understand their struggle and reality, and we need to see it as our own.

This is not a book on parenting, this is a book to help us understand who we are, why we make the choices we make and how we developed our mindset. It is a book in awareness of who we are and why we have become this generation of endless consumers and indecisive people who struggle, but try as hard they can to "get it right".

This book is filled with the insights I learned while being immersed in technology, helping famous people handle fame, and really digging into how our generation got here. We are the most privileged, and challenged in almost every topic; from relationships and parenting, to education, politics, and obviously technology!

My hope is that after you read this book, you can be more patient with yourself as you shape young and older minds, and more prepared to help them with what they may face. Hopefully too, we will enjoy and embrace these crazy waves of change that seem to come our way every day.

Introduction: Too Much Information

Everybody has an opinion about what you should do. We drive ourselves crazy trying to figure out in which position we should put our babies to sleep. Research shows that basically any and every position has its pros and cons. Formula or breast milk? Picking our babies up every time they cry, or letting them cry for a little bit? Public school or homeschooling? When should they start learning a new language? Which language should they learn?

Everyone and anyone trying to make an educated decision has two or more valid views on every topic. We are overwhelmed with information. We are overwhelmed with opportunities and dangers. While we have more information and resources than ever before, we are more frustrated trying to make good choices, and devastated when we feel we have made a wrong choice. There is an element of guilt and pride that makes every mistake feel bigger, and every win feel as if it were the least we could do.

Have you ever noticed that we are more defensive about our choices? When someone says something about our lifestyle, whether it is our personal habits, parenting, life, or career choices we just jump to defend why we do what we do. Why can't we just listen and let it go? We don't really have to make sure people understand why we made that choice. Think about it. Why do we feel this way?

We do not have to convince others that our choices are right. After all, it will be us who will deal with the failure and enjoy the success. Right? Yet, we feel like every choice has to be supported, documented, and proved to be right. We are the generation that cannot choose between being happy or right. We actually seem to only be happy when we are right. We have become obsessed with proving that what we do is in fact THE right thing to do.

We seek validation everywhere and anywhere, and we find disagreement offensive most of the time. There is something that pushes us to do what is "right". That is too much pressure for a human being in a world filled with so many options, is it not?

Now we are all also wrapped up in consumerism. It is not capitalism that failed here, it is the way we were raised, and the way marketing shaped our minds. While we cannot deny that we do live in a material world, like Madonna so plainly described it; we are still human and spiritual beings. We have more than mere utility value or purpose.

Now we are put in display from a very early age. In the good old days, our biggest embarrassment came when our family shared our baby pictures with our significant other or friends who came to visit. Now that happens online, and with the right settings, it can go viral.

There is too much information around us, there are too many opportunities and choices. If we are struggling finding the right path, making the right choices, imagine our children? Let us understand how we got here, to find better ways to get where we want to go. To find other ways to help other generations to figure out who they are and where they want to go.

This is why I'm writing this book. I want you to understand why we feel this way. I want to help you learn how technology shaped us, and how it is shaping others. We are caught in between a world where technology was revered and feared, and a world where it is simple and ordinary. The truth is technology is easy to use, but it isn't simple. It is part of our daily lives, but it is extraordinary. It should not be revered, or feared, but it should be respected and used with enough wisdom and caution. And it is likely that we are the only generations that can make that become a trend.

1. Fearless and fearful digital behavior.

The generational gap.

We have a variety of generations interacting in this planet. Some of which have witnessed World Wars, the invention of most of the tools we use today, and the tidal waves of change between the industrial revolution and the digital revolution.

Our current use of technology is thus divided into two main approaches that are mostly divided by the generation currently in their 30s:

a) Fearless digital behavior (under 30)
b) Fearful digital behavior (over 30)

Why is this phenomenon marked by such a generational divide? The 80s had a lot to do with our current love for technology. Those of us born and raised in the 80s were the first to find it normal to interact with computers. Before this time, computers were primarily used for work, not fun. Obviously, as with any generalization, there are exceptions. Some people outside of this "box" may have experienced the same, because they had older or younger siblings. Or perhaps, because they lived in different countries, with different technological and cultural influences.

This in-between generation has been the first to actually understand and enjoy the concept of videogames. We have witnessed and used each of our "cutting edge" inventions. From computers to AI. We are the ones that make franchises like *Star Wars*, *X-Men*, and the *Marvel* series, succeed. We are the ones who believed in science fiction as if it was a prediction, and not fantasy. We were actually expecting 'Back to the Future' and 'The Jetsons' to be our everyday reality when we got to the year 2000; and, by the way, we are still waiting for flying cars to be a thing. The generations before were perfectly productive and happy without remote controls, batteries, and video games. We, however were hooked, and ready for more.

So, let's start by talking about what is stopping us from embracing technology. We need to understand this first, since it comes from years and generations of experience and wisdom that are now mostly being kicked out of our culture when they should stay, at least partially. They are valid because they did come from solid life experiences. Most of the people who are "afraid" of technology and social media seem to share the following values:

- Work hard for what you want.
- Take care of your things.
- Never talk to strangers.

How do those values translate into our current approach to social media and technology?

Work hard for what you want.

Understanding the true value of work, resources, and the things you "get" meant that getting somewhere or something required a whole process. You had to know what you wanted, learn where and how you could get it, and then take a series of steps to make it happen. This whole process usually took a long time. In fact, everything we did in those times took hours, days, or even months.

Sending a letter would mean you had to:
1. Find a pen or pencil, paper, and envelope.
2. Write it.
3. Buy a stamp and put it on it.
4. Take it to the mailbox or post office.
5. Have the postal service collect and deliver it.

The process to send an email nowadays takes less than minutes, and it got cut to this:
1. Write it.
2. Send it.

Just like this process, most of our daily actions have been cut into minutes or seconds. Anywhere from communications to cooking, doing laundry, and even watching TV. Working hard has gone from a process that takes time, to almost immediate results. Of course, the response we get is equally fast. All of which turns a predictable and manageable way of life into a hectic, stressful, reaction-based way of life. We live each and every second reacting to stimuli and situations happening all around us.

The problem when you put the "work hard for what you want" mentality into our modern technology and communications, is that we would not expect solutions and procedures to be so quick and easy. Our brains keep looking for the mistake we made, the step we skipped, the "thing" we should be moving, pressing, or clicking. The challenge is to apply our thought processes in a world where you swipe, scroll, and touch in order to make things happen.

Take care of your things.

Phrases like "You break it, you buy it" and "Don't touch it! You're going to break it!" are drilled into these generations. Most things were hard to get, expensive, and hard to repair. Not everyone could have access to fix things. Not everyone could afford to get them fixed.

This feeling is particularly deep in generations that grew up in the midst of war or scarcity of any type. You only had one or two pairs of shoes. You probably had a handful of toys to play with. You actually cleaned your shoes and had them painted and repaired when needed. Household items lasted as long as they worked. This included cars, home appliances, tools, and clothes. Hand-me-downs were a regular practice in terms of clothes, and other items.

If you broke or lost something, you would not, and many times could not, get it replaced right away. You would have to work and save to pay for it, or wait until your parents worked enough to be able to afford it again. Many times, you could not afford it again or get it back.

As I mentioned before, in war zones, conflict survivors, just had different priorities. When you have strived to stay alive as your first priority, everything else may seem vain and superficial. It is hard to think about cyberbullying as a problem, when you've actually survived bombs and shootings. It is complicated to live in a world of updates and 2.0 when familiar was safe; when foreign means enemy; and when your mind is not used to seeing things as disposable or trivial.

This same mindset makes us extremely careful not to "break" things. And mistakes usually led to breaking things. Even in the first years of computers, pushing the wrong button would reset or delete most (if not all) your files. One command made a huge difference when you were trying to create a directory, run an operative system, or save your work. Nowadays, that whole reasoning and complexity is gone. Now you can click the right command or the right button and it is enough, and we even get a message that asks if we are "sure you want to exit/delete/open" something.

Never talk to strangers.

There used to be a time when children would go outside and play with their neighbors or on their own. They would only come back home when the sun went down, after playing outside for hours. Did you live that? Did you enjoy it? Knocking on your neighbor's door to check if your friend could come out and play? Asking for your other neighbor if you could walk into their yard or apartment to pick up your ball because it went inside their window? Or broke it?

In those times you would talk and interact with people you knew. They were your neighbors, your family, your friends. You were safe enough to say hello to them. You were expected to make a small conversation and keep up with them. You were encouraged to keep an eye on younger children, and keep them safe. You looked both ways before crossing the street, and sometimes you were able to "close" the street to play ball or ride your bike, or skate, or rollerblade… You probably delivered newspapers, or sold lemonade, or ran errands for family and neighbors. You were out in the world, but… Yes, there was a huge 'but'. The rule of gold in your childhood was: "Never talk to strangers"

How can you reconcile years of "never talk to strangers" to sharing your life and thoughts with the world? How can you work towards openness when there is a fear of the unknown from birth to adulthood? And for a very good reason!

Every time we fear the unknown it comes from the most basic instinct of survival. It was a system of parenting and socializing created to keep us safe, and to use communities and neighborhoods (those who were known and trusted) as the first line of defense of our young and our women. Many times, children and women were left alone, yet again, because "the man of the house" was out defending the country, or providing for the family.

Now let's talk about the other side. Fearless digital behavior. How did it happen?

The values shared by this community are:
- You can do it.
- You deserve it.
- You are special.

You can do it.

Empowerment. Plain and simple. It does not matter how many times you have to try, or what goes wrong along the way. You learn what works and you make it happen. Whatever it is you want to do, you can.

A generation of trial and error learned that if something did not work, all you needed to do was start over. A "reset" mentality perhaps. "If at first you don't succeed, try again." Nike's "Just do it!". Pop culture and influencers fully support this mindset. And do not get me wrong, it is not a bad mindset.

This is the mindset that produced the biggest technological development ever lived, and made Bill Gates and Steve Jobs turn failures into big successes. It allowed for great success stories like Michael Jordan, who proved he could "fly" in a basketball court. Or Oprah Winfrey who proved she could become the most influential and powerful woman after many people told her she couldn't and wouldn't be able to succeed.

Needless to say, this means that these generations won't mind touching, and moving, and looking for ways to fix, overcome or change something. This is probably the most tech-oriented skill we have developed as human beings. The opportunity to face challenges and succeed. Finding power in what you do, is very contained in our current technology. Even updates are part of this mindset. You do not have to get the software, equipment, or app ready and flawless. You actually know that evolution and use will make you face obstacles you may have never imagined. A system where you can update and fix those mistakes, while the user is already using the thing, allows you to think even if you failed at some point: you can do it!

You deserve it.

This is probably the most criticized element of thought of our younger generations. It is controversial, but it is also the result of our own evolution.

As opposed to previous generations, these younger generations, have been born into a world where they have basic human rights. From education, equality, and freedom to leisure, opinion, religion, and owning property. Many of these things, are not even universally achieved in most developing countries and underprivileged communities. However, most of the free world takes these and other concepts as a given, since their recognition in the Universal Declaration of Human Rights on 1948.

You can see that most social media and technological practices adhere to these principles. Everyone is entitled to an opinion, and to share personal views from sexual diversity to religious concepts. Most of us take that freedom as a given, and use it, or abuse it, without much concern or restraint.

Some education theories have promoted that there should not be any winners and losers in competitions, giving awards to everyone involved in a game or competition, just for their effort. Everyone deserves recognition.

Social media and technology are yet another thing we feel we deserve due to this mindset. We deserve to be part of the technological progress of society, and we deserve to be part of social media, and be safe and well-treated while using it. And some children think they "deserve" a cell phone or tablet. Do we deserve those? Do we deserve work and communications tools? Even when we are under the legal age to use them?

You are Special.

There are plenty of variations around this. You are unique. You are the best. Overall this means that you stand out. There is something about you that does not look or feel or sound like anyone else. This element of differentiation is probably the most characteristic within social media. Why?

We want recognition. As I stated before, we 'deserve' it. If I am special, then what I do/say/feel/think is special. And if it is so, then I have to share it. Or do I?

Yes, our uniqueness is the one thing that makes us so valuable when we are put together and brought into a community. We provide a unique history, set of values, and point of view. But, we are also very much like many other people. We share interests, perspectives, likes, dislikes, and many other things with plenty of other people. We walk into a full paradox where we are unique, and yet we find groups that share something in common with us. We strive to be found, loved, and followed due to our uniqueness. But, we are also trying to fit in and belong with others who may be similar to us.

The Best of Both Worlds

All of these ideas are deeply engraved in our minds. They shape our attitude and actions through technology, social media, and life. We go from fearless to fearful every now and then, and somewhere in between.

Using these resources fearlessly means that there is basically nothing that stops you. You may make mistakes, but you will go live, tag, record, share, and upload almost anything (if not everything). Selfies and hashtags are nothing to worry about. However, turning off your device, or disconnecting from the grid, are basically unthinkable.

Using those same resources fearfully means that we are always scared if something goes wrong. We may get stuck trying to solve it, and we will be overly critical of posting, creating, and sharing almost everything. We will see every interaction as a threat and danger, either to our property, or to our reputation, safety, etcetera.

Success and survival in this new era require a balance, coordination, and continuous sharing and learning from both sides. If you are part of the fearful generation, you need to embrace the newer values and learn how to apply them. Let go of the negative thoughts that come from your preexisting notions, but embrace the survival skills that you have learned.

The old values help all of us so that we can stay safe while guarding your privacy. They will keep us aware of the planet's needs and away from consumerism to use each device as best as possible. The hard-work ethic will also continue to be helpful and make us all resilient. Just because it didn't work the first time or it was hard, it does not mean you have to quit, give up, or disregard it.

The new values will help all of us understand our true value, know that there is room for our voice and presence in the online world, and find those special traits and qualities that help us stand out. That in itself is key for our personal branding process, something that in the past was just used by public figures; but today is essential to each and every one of us.

2. The Remote-Control phenomenon.

You can't always get what you want.
You can't always get what you want.
You can't always get what you want.
But if you try sometimes...
You get what you need.

The lyrics in this famous Rolling Stones song are filled with truth. That is, under a traditional and conservative view of life. Think of it as the "over 30" generations. But... There is a basic phenomenon that changed this: The massification of the remote control.

The remote-control changed our lives forever. It meant that we did not require physical interaction with that which we wanted to change. That in itself has a big impact in our lifestyles and mindsets. Even though the first remote controls appeared in the 1950s and 60s, it wasn't until the appearance of cable television and videocassette recorders, that things actually shifted.[1] And guess when that happened? Those wonderful 1980s.

---

[1] https://www.theatlantic.com/technology/archive/2014/08/how-the-remote-control-rewired-the-home/375214/ **Caetlin Benson-Allot, Aug 1, 2014, 2018.**

Remote controls were the stepping stones to wireless technology and devices. Beyond the technical derivations and procedures that took place, this actually happened in our minds; and continues to happen in the minds of our children during developmental stages. Most of the research on the impact of technology has focused on consumer patterns and comfort, particularly in work environments and home design. The most recent focus has switched perhaps to sedentary lifestyles and overweight problems.

The true impact of remote controls goes way beyond practicality. It actually means that we have 'control' over what we do, see, and hear. We have control over whether there is sound, brightness, and obviously display and contents of the device we are using. Channel surfing, switching channels during commercials, and recording our desired contents, came along with such device. But, they derive from the basic concept of 'controlling' our surroundings. This device just makes that quick and easy.

That same quest for control led to the evolution of our entertainment technology. As our options to choose what to watch, do, listen, or play grew; so did our capacity to store information. We developed devices that could record and hold information in a quick and easy way to store and use. We went from rolls to tapes, from tapes to disks, and eventually ended up with streaming services and cloud storage. All of this means, we have 'control' over what we use.

Yet... The model led to a mindset that is flawed. Where is the button that turns off bad news? Where is the 'mute' button in a child? How do we take off the batteries from a kid that just will not stop running around when we told him to? Here is where the remote-control users run into problems.

When I was in College (2000-2004) psychologists were starting to go from the X-Gen and Y-Gen to a new generation that was supposed to be known as the "playlist generation". The terms and segmentations have changed since. It is truly complicated to set a mark and a name for each generation with such disparities in evolution and trends worldwide. Many people refer to it now as millennials and iGens, or something along those lines.

I personally think that "playlist generation" was the most accurate description of what this up and coming generations would have as their main traits. Technology has given us choice and control, faster and easier than ever before. New generations have that from cradle to grave. As soon as they are born, they learn that the "key" to control their surroundings is hidden in those devices that grown-ups use all the time. They learn that if the click here, something happens there. Just as they learn that if they cry they will be comforted.

As I was stating before... What happens when we encounter those things which we cannot control? How do we learn to cope with that? When you come from a generation that grew up facing trials and tribulations without being able to "shut them off" fast and easy, you have more emotional, psychological, and even social tools to adapt. You know you have to wait, or fix, or ask for help. But we have lost this, and it is precisely what we need to pass along and build in newer generations. They will not have it or find it on their own, because their environment, and their communities are not demanding it from them.

The big problem here is that we are losing those skills. We, ourselves, are losing our patience, our ability to fix, and to ask for help and support. In the few or many years in which we have interacted closely with technology, we have become dependent. We need control.

We want children who are quiet, and organized, and obedient. However, those skills are learned through many hours of tantrums, mess, arguments, crying, and screaming. You may get lucky, and have a baby with a mild character that will stay quiet and learn fast and easy. But, with a universe of overstimulation, that is most likely impossible.

Control your mind and emotions, even when you cannot control what happens.

You cannot and should not control everything that surrounds you, especially when it comes to people. You cannot control the weather, traffic, and many other things because there are things beyond your control. And even if you could control them, millions of others, just like you are trying to get things to go their way.

Emotional intelligence, spirituality, mindfulness, science, and education overall are key to the self-control process. Control your thoughts, your emotions, your reactions, and your actions. Understand that there are things beyond your control. Understand how the world works, how people behave, and how you can and should react to that in different scenarios. You cannot and should not do everything and anything you want or feel like doing.

Now, be patient with yourself. Chances are your will to control it all will grow desperate when you notice that handling your emotions does not mean suppressing them. Here again, is where we have gotten it wrong. So sadly, and regrettably wrong, that it has cost us the lives of many of our most inspiring icons, and many marvelous human beings.

In our attempt to control everything, including our emotions, we thought "not feeling" was controlling our emotions. We have looked for that "not feeling" button in the bottom of every bottle, the first to last cigarette, and millions of pills. We have disregarded feelings as "complicated" and filled ourselves with casual relationships and pleasures to avoid complications. Only to find that no matter how far we go, thoughts and emotions will follow, and they will not be shut down.

So again, don't confuse control with suppression. Go for management. The ability to focus our resources where they are needed. Our joy and sadness have a cause and may have positive or negative outcome. That is precisely what we need to learn. All emotions are needed, we just need to find where and when.

See just how right the Rolling Stones had it... Sometimes you get what you need. You get the opportunity to step out from a world of plugs, buttons, and programming to being human. When things do not go out was you want or expect, you may find exactly the lesson or opportunity you needed.

3. The Frankenstein Phenomenon.

I think we are all familiar with the story of Frankenstein by Mary W. Shelley. In summary, Dr. Frankenstein creates a being from body parts, because he finally found how to do it. After that, he faces the consequences of what the 'monster' does, because he never expected him having a mind of his own.

This same phenomenon has taken place since the beginning of science. Scientists strive to find why things work or happen, and they run experiments. They have a notion of what may happen (a.k.a. theories), and they try to see if they are right or wrong. I know, and fully acknowledge that this is a very very simplistic version of the scientific method. I do know I skipped plenty of other steps. But, this is to show that most of the time scientists, and innovators overall, do not know what will happen if and when they succeed. They dread failure, but they rarely plan for success. The case may also be made for entrepreneurs. We may all do our best, but we do not know exactly the impact of our success.

Why does this matter? I would say now more than ever before. Well, because now we know more than ever before. Those who created the first engines, devices, and tools, did it to solve a problem. One main problem. Many times, disregarding or ignoring things that could be related to that problem. For example, Marie Curie discovered radiation, without knowing how dangerous radiation would be on her body. People started to use coal as a source of power, and did not think about the health or pollution consequences. First because they did not know them, then because they did not have an alternative.

The problem is that our innovators keep making the same mistakes. Mark Zuckerberg created and grew Facebook without acknowledging the social, legal, political, and international impact that it would have. So, he was unable to detect, plan, and control for it. Is he responsible for his monster? Did he know what would happen? Could he have known? Could others have known and planned for it?

We cannot and should not attempt to stop innovation. We are meant to evolve, and a part of evolution is making better things and doing things better. So how do we stop the Frankenstein phenomenon from continuing a chain of chaos and world destruction? How do we tame the monster?

First, we need as much multi-disciplinary education in our lives as possible. We need to know or see a problem from different angles. We need to assess the impact of our actions on everyone that may be affected by them. That in itself means we need to know who may be affected by them. Who are they? Where do they come from? Granted, that may be a stretch, since, as a social scientist, I do know we could never measure the true impact of all of our actions. But let us try an example.

If I am developing a new software that will do something, anything, I need to know:

a) What I want it to do.

b) How I think it will do it.

c) How it may be affected or affect other components of the machine, its software or hardware.

d) How it will be operated by the user.

e) A general vision of the things that may go wrong if it does not work properly.

The case can be made for everything and anything we do. It can be a business, a project, or simply our daily routines. We can no longer just choose based on our needs, our desires, or our capacities.

Here is the lesson:

I am responsible for what I do.

This is probably one of the most painful lessons we learn as human beings. When we hurt someone, or we hurt ourselves. When we break or lose something we love. When we lose that precious control we cherish so much, and something goes wrong. But like I said, now we need to assess the same things for when it goes right.

As an entrepreneur, what will happen if I succeed? How will I treat my employees? What are my values? Who will I work with? If

I keep growing… What will be my priorities? Do I value profit over quality? Do I know the social, economic, and environmental impact of my actions?

These sorts of questions were left to managers and CEOs traditionally. However, as the economic and work markets evolve, entrepreneurship and freelancing force all of us to understand and manage this too. We can no longer expect others to know what we should do, when we are working in schemes where we are in control. Oh the joys of freedom! Oh the wonders of being your own boss!

4. It takes a village.

Who raised you? Who took care of you when you were growing up? Where did you sleep? I have noticed that we have forgotten how many times someone helped mom or dad. Our memories are usually of what we did and where we went, but we don't know why.

Our recollection of our childhood skips all those times someone else took care of us because mom and dad had something to do, somewhere to be that wasn't with us. We may remember the things our parents missed if they were important dates or events, but we may not have any idea of the time they left us at a playdate, or relative's house, while they just got the car fixed, or did something of their own.

One single person has never been able to raise a human being. Even in *The Jungle Book* and *Tarzan*, you see that a whole pack of wolves, bears, or even a whole number of different animals raised the child.

And yet, for some unexplainable reason, we keep acting and talking as if we were the only ones raising our sons and daughters. We tend to think about what mom and dad have to do to raise a happy and healthy kid. We use all this information that we have gathered throughout the years, and we add up what everybody knows in online communities, groups, and books. Then we use all of that to try to figure out what we will do.

Then, all hell is loose when we realize we cannot do it all on our own. When we notice that we are fully in charge, panic invades us because suddenly everything we do is critical. The truth is we cannot do anything on our own. Our child will have millions of influences, many of which we cannot control. We could attempt to reduce the impact, but where do we stop? Siblings, relatives, friends, neighbors, school...

Honestly, parents and care-takers are essential, because they will always be the first line of defense. We are the first line of defense, but we cannot control exactly who our children will become. Just as we did not become who our parents expected, for better or for worse. Our generation has witnessed a complete differentiation in traditions and habits from the generations of our parents and grandparents. We are closer to our parents and children. We have more freedom and choices. We are more diverse, and more inclusive. And now finally, our tribe has gone global.

Regardless of where you live and who you are this is true, and will continue to be so. The only way to make this work for the betterment of future generations, is to get them ready to make their own choices. They will have to choose, based on their values and beliefs, who is going to be part of that tribe that influences them. What shows, artists, public figures, and influencers overall will become part of their tribe? Who will shape them?

The actions and choices of every individual matter.

As a global tribe. All of what we do and create influences others. Every content we post online, every action we take offline, everything counts. It counts in economics and markets, in our business life, global warming or sustainability, and our relationships. The butterfly effect is a reality online and offline.

5. Keep it to yourself.

We are fragile creatures. Even when we disguise ourselves with indifference and superficiality some times. What other people think and say about us does make a difference. Regardless of how much we work on our self-concept and self-esteem, we care about others and how they view us.

When we go online, we are seen. Younger generations have yet to understand what that means. They are not afraid to post everything they think and do online. And they live boldly mainly because they have not yet understood the consequences of their actions.

People on-screen and off-screen

Social media and the digital era have brought another big challenge into putting people first. Screens seem to dehumanize us. We see the world, and reality, behind screens and suddenly things that would shock us seem normal. Violence, destruction, and pain are normalized. Sex is normalized. And again, do not get me wrong, I am not saying that sex is "bad", but the severe lack of sensitivity into what human contact may bring has made sex trivial, when it is nothing but trivial.

We all need to remember that we do not stop being people when we are behind a screen. We have lost track of things because entertainment changed in format. We used to think that things on-screen were part of a make-believe world. As with children, we learned as time went by that the news were real, documentaries were real, and we could filter what we saw. We learned which channels were portraying reality and which were portraying fiction.

Reality TV and the growth of self-generated content have led to a 24-7 mix of desensitization. Everything seems like part of your entertainment content. The video that your friend uploaded, the news, and that funny video by your favorite influencer. What if they all shared the same story but each gave you different information? Who would you trust? Who do you think your children would trust? If they see violence... Do they feel scared? Or do they see it as another action scene? If they see someone talking about their personal life, do they find it entertaining? Common?

We are our kid's guides. We need to tell them that just because something is on the other side of the screen, it does not mean it does not affect them. We need to show them the difference between reality and make-believe. And we ourselves, need to remember that we need to question what we see. Not every ounce of content is to be trusted. Not every post is filled with truth.

These days, we get all our information and contents online. Seeing our shows, right next to a news clip, and an update from friends and loved ones plays with our minds. Everything tends to blend into either fiction or reality. It is hard to keep up with things! It gets complicated for our minds to know what is real and what is not. Misinformation thrives on this phenomenon.

There are people who take advantage of this. Especially in young minds. After all, remember, older minds like ourselves lived a different formative experience. We had an opportunity to learn how to filter information. We learned how to trust and validate content, and even chose our social references and trusted sources way before we were overwhelmed. New generations are overwhelmed with content, and have not had a chance to learn who they should trust.

We all matter, we all deserve respect, and we all make mistakes.

What you say online is seen and heard by the world.

I emphasize this, because screens give us a fake sense of privacy. Even though we know that we are interacting with millions of people around the world, we usually post from very private settings. All we see consciously is our interaction with a screen, a keyboard, or a device that seems very private, but it is not.

My general rule is… What would your grandmother think of what you are sharing? Would you still share it? It applies to posts, to comments, or anything you share. It is not just about keeping things great with grandma. It is just a general rule. You can also think What would you do if your boss was reading or seeing this? Because few people are aware that the opinions they express in their personal social media profiles, are actually seen and checked by people they work with.

This is yet another reason why you cannot split your life into separate areas or circles anymore. Privacy, while not impossible, is very unlikely to be kept. Mostly because people share too much, and they do not keep things to themselves.

This is why you have to be aware of your tribe, and the tribes you may belong to as well. The new digital age allows us to be many things and not fall just in one box or category. However, this means that the complexity we have as human beings is more evident and we need to be aware of that. Our complexity does not make us invisible in different settings to different people. Our complexity allows for bigger and better connections. It is empowering, as long as we are aware of it. If we know who we are and where we stand we are better equipped for our digital interactions.

Trolls

I am sure you have heard this word a lot, and I am confident you are not entirely sure how they work. Many people confuse trolls with bots, and there is a huge difference.

First let me explain why I am devoting an entire subsection to this. Just as all of our traumas and behavioral problems came into social media when we all migrated online; all of our toxic and negative social media behaviors are migrating to our offline lives.

People who did not have boundaries in real life, started oversharing online. Naked pictures, sexting, and the abuse of selfies are a consequence of this. People who were insecure offline became "haters" and started a series of behaviors and habits that are now known as trolling. Bad behavior offline led to problematic digital behaviors, but now the trends go both ways.

Oversharing and a lack of boundaries is overflowing in real life. Our children and teenagers see behaviors online that they copy in real life. Online influencers are more powerful than real-life influencers, because we are more exposed to those online than to those offline. How much time do you spend interacting with people? How much time do you spend working on a screen?

The sad truth is that trolls start appearing in your social media platforms when you start succeeding and growing. They are like weeds that appear in your garden when you have plenty of water. These people just spend their time finding accounts that are successful to use the platform to be seen. They thrive on controversy and scandal. They will try to make you respond to them and engage. If you do not like, share, or comment a post, it "dies" in the algorithms. But, as soon as someone engages with the content, it is brought to life. Trolls know this. They will do, say, and post, whatever they have to, in order to get you, and your tribe, to engage.

By now all of us know what bullying is about. Someone picking on you because they feel stronger or more powerful. They know you and they do their best to make your life impossible. Your big brother, that kid in school, or the neighborhood bully. You know who they are, and they know you.

Now imagine you are walking on the sidewalk, and someone you have never seen comes out of nowhere and pushes you. Not accidentally, they actually just walk in your direction just to push you because they can. What would you do? As a rational human being, you may ask yourself: Why?

This main difference between regular bullying and trolls is that one has a reason for attacking you. Bullying is constant aggression against you. Think of it again as if you ran into someone you know and they pushed you in the streets. You know who they are, they know who you are, and they want to push you around and impose themselves, because of the relationship or environment they share with you.

But trolls go beyond, because there is no explanation to their behavior or harassment. They do not target specific topics, they go after all and every comment that can stir people's emotions. You might go insane trying to figure out why they are bullying you and intimidating your tribe. Trolls do this, just because they can, and they felt like it at the moment.

This is the most troubling phenomenon that has started jumping out of our screens and into the streets. Nowadays, you find real-life trolls. People who just create chaos to get attention. They even record their aggressions and make them viral. They just thrive on chaos, and when they do not find it naturally, they make it happen. They will yell across the street and become violent just because they can.

Sometimes it is a mix of both. You probably have heard about misinformation tactics. Misinformation is content created to appear to be true, when it is false. People in an office, for pay or pleasure, actually generate this content. They take a photo or video and edit it to fit what they want to share as news. They may say that there was a fire, or an accident, or that certain politician or public figure said or did something that they didn't. Sometimes they do it as part of a political campaign. Sometimes they just do it for fun. They love seeing their content go viral and shape real-life events. This is how trolls mix both platforms. They create chaos in your information, so that your real-life actions and reactions have real-life impact.

So how do you deal with them? Online, you block and report them. Even if you do not find them in your own timeline. Wherever you see them, stop them right away. Do not reply to their comments. Do not show other people that they are there. Do not make them more visible or relevant.

How do you deal with them offline? Again, you go to whatever local authority you may find and you report them. They love attention. They love it if you react or engage. Walk away from them and find a safe space where other people can stand up for you.

As a general rule, if you have a real gut reaction to something you read or see online. Try to look for more sources, talk to people who may be closer to the problem or who may know about that specific topic before reacting or sharing it. Double-check things, and at least watch or read full contents before you share anything online. Users are the first line of defense against misinformation and trolls. Authorities take too long to find them or react.

It takes a village.

Not only to raise our children, but to keep all of us safe. The creators and developers of most digital resources online know that there is no system that can keep all of us safe. Furthermore, they know that there is no system to control the problems that derive from their platforms and resources.

Have you heard about hackathons? Did you know companies actually give rewards to those who can show them flaws in their systems? This is meant to overcome human errors and making communities and systems safer. This allows for different people to look at a problem and figure it out.

When you are using social media, the strategy is the same. Platforms develop tools so that you can report abuse, violence, or misconducts overall. The systems are not perfect. I have experienced a couple of situations where Facebook and their support systems were completely useless to stop violence. Local authorities were unreachable. Only organized users were able to make things better.

One time, someone had set up a Facebook account where they were portraying illicit images of under-age users. These were photos and videos of children in their teenage years in very sensual positions. This was very close to child pornography, but it was not. It took me, and thousands of other people, reporting the pages to get Facebook to react in more than three days. The main problem was there was no specific option on Facebook at the time to report the specific incident. Like I said, it was not child pornography. It was not violating copyright laws. It was not aggressive against me or someone I knew. And there was no way to get in touch with a human being to explain the situation.

Another time, people started sharing the photo of one person under the identity of another one, and shared a message without the true identity of the author. This message started spreading online in less than 24 hours. It was clearly being used by a collective to promote a message and promote chaos. The person that had originally posted the content started receiving threats. The message kept spreading, taking different turns along the way, violence and chaos kept going. And regardless of how many times we reported the message, new accounts started using it, some were fake, and some were real. Every now and then I still run into that message, it still is attributed to the wrong person, and no one has been able to stop it.

The reason all of this happens is actually freedom and diversity. We are all free to share contents online, regardless of their validity. We are all capable of downloading the image of another person and creating a story about them. Who can stop this? Other people who are aware of what is going on. There is no regulation that would be fair. There is no platform that can prevent it. Just people who team up and help each other. Some by stopping those who share the content, and some by protecting themselves and the people who created it. Otherwise we run a risk to just fall into general censorship where nothing and no one can talk freely about what they experience, and that ruins the whole of social media.

The same thing will happen with the human experience for future generations. Full trust cannot be placed on any of our influencers or guides. Our own judgement is the best tool we may have, and developing it is the biggest challenge, and it will continue to take a village.

6. Bonus points and reset buttons.

I have played video games for as long as I have consciousness. Maybe a couple of years after that to be accurate. I started playing videogames on a computer, then an Atari console, and later on in my Game Boy. Games actually came in floppy disks, and you were beyond privileged to have one of those copies. I loved the thrill, the colors, the music. The tunes for "Tetris" and "Digger are still engraved in my mind. Try finding them on Google…

I am one of those people who actually felt deep emotions when *Nintendo* re-released their classic console, and Adam Sandler released *Pixels*. I jumped with Mario, and took turns with my cousins and brother so that we could actually finish the hardest levels. As an adult I fully enjoyed playing *Guitar Hero*, and dancing with games and consoles that detect your movement without having to code the A+B plus the other button on the top.

I love video games, and I have a serious problem with them. I get hooked. My brain is wired that way. That is why I stay as far away from them as I can. I do not download them to my phone. I may have some of them in my iPad just to prove that I can use them wisely now and then. But I am aware of the addicting elements that are included in them. Early on I learned that the music did something to my brain. I just HATED hearing the sounds and music that made me feel like I had lost or failed.

I learned to play videogames socially, for the most, so that we could all hold each other accountable when enough was enough. Later on, systems started allowing us to interact with each other while we played, even if we were miles away from each other. You could even ask for help and collaborate in each other's games. If we were playing Farmville the collective could empower you to keep going and growing and achieving your goals. Way to go utopic! Right?

Except, those games that were "safe" in our days, because you could stay home and stay out of trouble, were not anymore. Aside from the obsession with playing "just one more" till you won. Aside from the effects of colors and sounds in our brains. Aside from the motor problems caused by the constant repetition of movements in a joystick or control.

The true danger now lives within the online access that our gaming consoles have. Nothing can keep us safe once we go online, other than our own choices. And trust me, nothing. There is no filter, no code, no parental control, or anything that can stop someone from getting into another device. If there is, someone is already working on the way to break it or bypass it.

Nowadays when a kid has access to a gaming console, they are free game for other users to contact them. Our children are not safe inside their own homes. Did you ever feel that way growing up?

The insensitivity of video games.

Many studies have been conducted to prove and disprove how video games—and action films for that matter— have made us more prone to shootings or violence. On the one hand they are venting options to discharge our frustration and aggression without hurting others. On the other, we get used to killing things and witnessing violence without reacting accordingly. Whether we stomp on the mushrooms as *Mario*, run away from the ghosts in *Pacman* or take full control of a war operation in *Call of Duty*. We "kill" others and we get points.

Something else happens on video games that affects our view of the world. When we die, we do not die. When we kill and see people being killed we do not even flinch, actually we are "excited" and thrilled to see it.

In the old times, in videogames you could actually feel disappointed because you lost all lives and could not go on playing (unless you hit reset obviously). Nowadays, you can actually buy lives. You do not even have to wait or take a break until you can try again. Think about it. What does this tell your brain about life and death? Regardless of your values and formation, repetition trains your brain.

I know many people who have balanced this in a wonderful way. People who love videogames, and also tried martial arts or boxing, or playing sports. That way, they actually balance the lack of reality of pain and physical response with a real body experience. No matter how much they see that online, or in a screen, punching does not hurt, the bruises in their bodies show them differently.

In the midst of all this, there are tools that may be really good and positive from our online lives. Dealing with games and devices for years, I learned that pushing the reset button was liberating. It actually allowed for bugs and problems to go away, and devices to work properly. Sometimes there is just too much going on behind the screens in any operating system, including our own.

In life as in tech RESET solves most of your problems.

In our human experience, "resetting" our systems is good. Stop all actions and start from zero. Your approach changes, you can see things clearly, and you find the answers you were looking for.

There is a certain logic that springs from technology and gaming that is actually good. Things like team work, reasoning, strategies, mathematics, engineering, problem solving are helpful. It is not all bad and addicting. You can find some life truths in games, because once again, it was humans who programmed and coded them. There are humans behind and in front of the screen, always.

We need to sensitize ourselves back.

It is not only video games, it is our whole culture of watching life through screens, that has desensitized us. And we need to reconnect. We need to feel something when we see pain. We need to get angry when we see something unfair happening.

Some people like libertarian Gloria Alvarez, criticize millennials, and I would include our generation, calling us "Starbucks activists". What she means is that we all complain about things happening, but do nothing about them. She talks particularly about people who complain about how capitalism is the devil, while sipping on their hipster brands and trendy coffeehouses, obviously from their latest iPhone. They deal with world problems from the comfort of their safely kept lives. They think posting a problem, automatically solves it. The most they will do about it is tweet, click, and complain.

I agree with her in many things, but not on this. To me, if people are at least feeling something strong enough to allow these topics to be part of their timeline, there is still some hope. You can do a lot with a click and a post these days. You can do a lot by sharing the right information. You can donate by just clicking and give money and resources to organizations that are actually doing something! So, it is not so bad if all you do is "click". Just as long, as you are clicking in the right direction, and not just whip and nae naeing.

We need to remember that life is not about controlling our environment, and shutting out humanity. We cannot pause our lives, mute those who think differently from ourselves, and block people who hurt us. We need to face them. We need to live life. We will suffer, it will be hard, but lessons must be learned through each and every struggle. We can use the hard times to learn how to ask for help, practice our team-work, or use all that knowledge that we have gained to serve, connect, and heal. We can use hard times to understand just how good we have it in the good times. But mostly, we just need to learn who we are in those moments.

This is why our children need to be brought back to feeling. They need to feel pain to know how to handle it and focus it in a positive way. They need to feel fear to know how to react. This does not mean you just leave your kid in the middle of the street and walk away just to see what they will do. Or that you will have them watching the news all day. But, you can guide them through scenarios in movies and books and use those situations to help them cope with things. Then you can go a step up and have them see the news with your supervision.

Remember fairy tales? Those were and still are great stories to use for good and deep conversations with your children, not just to have them dress up as Ana or Elsa and sing "Let it Go".

There is also the natural opportunity of having them take part in some "adult conversations". I remember many family and friend gatherings with my parents when I was growing up. While there were "adult conversations" that we were not invited to, we would be encouraged and welcome to join some of them. I learned a lot. I sat through topics from traditions to banking. Now, I appreciate it much more. Had I just been welcome to talk about games and school, and not listen to others, I would not have the insight and experience to connect generations. I would only know history from history books, and my perceptions would be terribly biased.

Fake news has been a thing for a long time, and the only way around it is to go to the sources. That allows you to learn about emotions and consequences, not just facts. This is the kind of sensitization that we need now. We need to talk to people and learn from people and their experience, and tie it to the facts we learn. What better platform could we have for that than social media? We can actually talk to people who live what we only read about. They are our sources and most accurate fact-check.

7. More than words.

Words have power. Words have meaning. Our language is important, because it allows us to understand other human beings. Even non-verbal communication is essential. It does not matter if it is online or offline. Emojis matter too, you cannot use any random sign and expect others to "get the right message".

Let me share a story with you. Months ago, I was visiting Miami, and had to take the bus to get from my hotel to the hotel where the convention took place. It was a long ride, and I usually take a book and check my digital to-do's to occupy my time. But every now and then, I remember I actually enjoy watching people. I may just keep my headphones on (yes, I am part of the Walk-Man generation, and I still use my iPod and not my phone for this) and enjoy watching how people interact and behave.

I am always amazed at the behavior of teenagers, but the boy I saw this time was maybe ten years-old. He was riding the bus on his own, just out of school, and was holding a conversation with someone. Up to that point, it was not that extraordinary. What I couldn't believe, was that he was on a videocall, and he was not even looking at the screen. And get this... The guy on the other side of the screen wasn't either. Why would you hold a videocall when you have no intention of seeing the other person?

I even struggle when I'm on a videoconference and the settings on my computer or the software do not allow me to make eye contact. Eye-contact is key! Did anyone else get called out while growing up for not looking at grown-ups when they were talking to you? This is why! Because eye-contact tells us we are listening to what is being said.

Our body talks and we need to learn what it says.

We learn to communicate from a very early age. I have collaborated with psychologists and parenting experts over the years, because our children are not developing this awareness. The exceeding amount of screen-time that young children have is stopping them from learning about human interactions, communication and emotion. It is not the same to learn from characters on a screen, than it is to get to know your family and caretakers. We copy everything we see from the day we are born. And I mean everything and anything.

After we understand that our gestures mean something, and crying gets us a bottle, or a blanket, or a clean diaper; we learn words. We learn who is mom, or dada, or all those other cute nicknames that we develop when we are trying to communicate. Sounds put together mean something. And the way we use those sounds, their order, and their frequency makes a difference.

Once we get them down, we learn the rules, structure, grammar, spelling, and all those fun things that get thrown out the window when children in their formative years start using random letters and typos. We all did it growing up. But, what percentage of our communication happened through initials and emojis? What percentage takes place now?

Trust me, I have plenty of fun dis-arming not one, but two, languages. I love making up terms, and putting one language into the other, just because I can. However, I did it after I had learned the rules and abided by them for years! The real problem is when you do not get the chance to learn them and put them to practice properly. You may learn the importance of good communication the hard way when you are trying to read a contract or listen to the charges pressed against you. Yes, that is how bad things may get if we just neglect the importance of words.

It is not just a problem for business or formal settings. If we cannot use words, we cannot describe how or what we feel. We cannot engage in healthy relationships and explain what is wrong. We cannot have a smart dialogue with those who may disagree with us. And what happens? We run away, we quit! A simple problem that could have been solved by talking it over, turns into a breaking point situation, because we cannot communicate. Relationships, families, and jobs are thrown out the window.

We need to keep communication up and running. It is essential for all of us as a global community, and it does not diminish with technology. If we cannot get our point across, in any language, to people around the world, we are lost. We need to write emails, and send texts, and give speeches or even record YouTube videos that make sense. We cannot go backwards and record everything that happens so that we can share it so that all we have to say about it is "watch".

Even miscommunication is useful! I work with people from many different countries and backgrounds, and use English as our primary language for communicating. Even the same words and expressions have very different impact in each of us. Sometimes written communication is not enough. Sometimes we need to clarify what we understood from a very simple message. Sometimes when I work with customers developing content or websites, very often we need to clarify what "more color" means. Or why I decided to change a button for a link or vice versa.

Defend communication in your life, do not avoid it, even when it is uncomfortable. Ask your children how their day was. Ask your spouse how they feel. Talk to your friends about the latest movie or the political climate, but keep talking! Check your emails and reports to make sure you are using the right words.

It is for your sake, and for your children's sake. The rules of this game called life are changing all the time. What works today may not work tomorrow, but the only way of learning about it and helping each other, is constantly communicating.

8. TTYL, BFFs, and Tinder.

What would we be without relationships? If your answer was "happy", I have to say I agree with you, but that is a problem. We are social creatures. Even if we make up for it by socializing with dogs. We are disconnecting ourselves from people in our lives so incredibly often, while at the same time fighting to connect with people online to "grow our communities".

We keep people at a safe distance. Close enough so that we feel connected and far enough so we can disconnect and keep our own space. The next generation went one step further, they barely make an effort to be with people. All they have to do is turn something on.

TTYL

TTYL (Talk To You Later) are the now famous initials that tell us a conversation is over. Or in a different scenario, that we are intruding on something we cannot hear.

Do you remember the thrill of calling your friend's house? What if their mom picked up? One way or another all of those filters built us. They shaped our personality, our social skills, communication and problem-solving strategies. Each and every struggle in our lives has helped us build those strengths.

What filters do you have nowadays for your children? Do you meet their friends? Do you know what they like and share?

There is obviously a sense of privacy and independence as we grow up. From our rooms and lockers to our conversations, there are things our parents and grown-ups may have never known we did or said. How about now? How scary is it to think your children are holding those conversations with complete strangers?

We do not want our children to cut the conversation as I mentioned before. We do not want children to have to "hide things" from us. We want to keep them communicating, sharing, and making us part of who they are and what they want. We may not be able to monitor or control all we want in their lives, especially online, but we can stay part of them. An active part hopefully.

BFFs

Most of our friendships and relationships flow and remain through social media. How close can that be? What kind of relationships do we build online? This gets tricky because social media allows us to keep a safe distance. Oddly enough that same distance makes us more comfortable in sharing very intimate information about ourselves. It enables us to do and say things we may not have said face to face.

In a way people share more and more of what they feel and think than what they do with their friends. We may share videos and photos of what we did, but we do not socialize as much. People know about what we do, but mostly witnessing through what we share about it. There are less memories in our minds, and more reminders of "you posted this a year ago".

Do you show your children how to "be a friend"? Do they see you communicate or talk about your friends? That is the best lesson they can learn about friendship, and relationships overall. They learn more from what they see you do, than what you tell them to do. If all your interactions with friends are social media posts and chats, they will learn that that is what friendship is about. If they see that it is just a portion of your relationship, they will learn that too.

Influencers

We cannot talk about online relationships and relationships as a whole, without talking about influencers.

People who share your values or worldview, online or offline, become your influencers.

This has now become a paid job, because more and more people rely on what other regular people do and promote; even over traditional information sources such as the media and public figures.

Brands are aware of this, so they set budgets for influencers of different areas of interest and impact. Micro-influencers (people with thousands of loyal followers) have been known to be more efficient for some brands, than those with millions of followers.

How do you become an influencer? It is as hard or easy as building a strong base of followers. You have to gain credibility and people who actually care for what you do or say. Sometimes this means that you build a following just because you are funny, or maybe because you look good. Yes, that is how easy that works. Sometimes it is as hard as actually working on your personal brand and supporting the things you find useful and valuable. Remember those good old values we talked about?

When I worked with up and coming artists it was hard to convince young people that fame was not always a good thing. Sometimes they learned this lesson the hard way. They lost privacy, or their mistakes were made public, and they were not as happy to be seen and judged by thousands or millions of people. This same thing happens now with any and all of our youth, as well as ourselves. Building and having a following means that people are looking at what we do and, unavoidably, they will have an opinion.

One would hope, that the golden rule of "Do unto others as you would have them do unto you" would grow with this, right? If we are all learning what it is like to be seen and judged, maybe we could be kinder and gentler on each other? Sadly, this is not the case. As social media grows, we all feel like we are entitled to an opinion and to be the judges of what others say and do. Yes, social media gives us a voice, a strong one. It is our choice how we use it. And this is how we can change the way social media works.

Tinder

I do not intend to promote a specific dating app or guide you in how to use it. That being said, if you know nothing about Tinder, this is a way to find people you may like and want to connect with (there are options, messaging to follow-up or meeting). How do you select them? You see their photo and swipe to the left if you do not like them, or to the right if you do.

How many of your friends would you have become friends with just by seeing a photo? How many of the people you dated or had relationships with would still make the list?

Everything in our lives has become instant decision-making. Not to mention that there are other apps that are specifically designed to get you sexual encounters. Nothing says human connection like finding each other in an app, having sex, and never knowing about each other again. Right? I really hope you do not agree with that at all!

It is not that random hook-ups did not happen before having a smartphone. There is always an option to keep life and relationships as casual as you may want them. Screens make it safer and easier in many ways. But we go from superficiality as an exception, to having it become the rule.

Our minds and souls are so confused nowadays about how to find connection. We are judged by a good or bad photo. We are judged by what our profiles say of us. And we are so busy working (most likely in or with machines), that we rely on social media to keep us socializing.

9. Guarding your soul in a world of likes.

Life has sadly become an everlasting non-stop popularity contest. Your self-esteem seems directly connected to how many people engage with you online. I secretly do feel a little sad when my posts go unnoticed, even though I know it's more the algorithm, or habits, and not the reflection of my loved one's true love or connection to me.

How do you survive this? How do you build a healthy self-esteem and appreciation? How do you keep good relationships in the age when you think you KNOW if people are avoiding you?

Define your values

Let's talk values… I'm talking about your core values, those that make you who you are. What is important to you as a person? What matters to you as part of a community? What makes you get up every morning? What do you pray for every night? What are your priorities?

You know what they are? Great, don't forget them! Every single thing you do MUST include, reflect, and carry your values.

The hard part of this, is that, as we have mentioned before, human beings are not fixed and unchangeable. We change, we evolve, we have a mind of our own and traumas and things that inspire us. So, once we get to "know" ourselves, we realize that we are not really what we thought we were. Even as we know ourselves, we notice how much we change and adapt.

Selling yourself

We already mentioned we are becoming brands, and basically, we have to sell ourselves. At first it may just be to your friends and family. Let them know what they can find in you. Let them know what your role in the family dynamic may be.

This is not just about fitting general stereotypes, of the "funny one" or the one who is "good at sports". Think bigger. Who do you go to in your family when you are starting a project? Who do you approach for advice? Who do you look for when you need a different perspective on something? Maybe someone who is great building, or someone who is amazing decorating… Your roles will change in time. What you can bring to a family as a kid, changes fully as you grow into your teenage years and adulthood.

Stepping out of your first social unit, you need to learn how you fit into society. Your church, clubs, school, friends… What do you bring to them? Again, it cannot always be the same. You will adapt and evolve, both within yourself, and within your social units. You may bring something to your family that someone else brings to your friend group. You have an opportunity to have a complex personality! You have an opportunity to thrive differently in different settings! You can be an artist and an athlete. You can be spiritual and practical. You can be funny and profound. Break the mold!

If you learn this. You can win at this game called life. Once you actually have to make a living, and go to the ultimate level of "selling yourself". When you actually get paid for being you. You will find it easier to fit into whatever work setting you take. Even when you do not go into your dream job, you may find a way to be happy, successful, and productive. Why? Because of years of being comfortable finding a setting where each of your strengths and opportunities are fulfilled.

Our strongest suit.

Our biggest asset is our humanity. The values, compassion, and connections that we may develop, will remain our strongest suit.

I am not going to lie, when I talk to my developer and programming friends I get chills. We are moving so fast and beyond any control with technology! While many of us are trying to get the hang of what social media is about and how we can use it without going nuts, there are plenty of new tools and technologies that we have not even heard before. Many of them, will become as daily and conventional as our smartphones in the next three to five years. But the truth is, regardless of what technology does, what matters is what we do with it.

Every new technology has been received with some fear. People were unsure about electricity, telephones, cars, and almost any innovation that we have had. Again, some of that fear was good. Some of that fear, when developers actually listened, could be used to avoid or reduce negative consequences. Anywhere from seatbelts to passwords, there are many things that can make technology safe. But nothing is safer than when you have educated users.

The best way to succeed and sell ourselves, is to be and know ourselves.

That is the big trick. That is the big secret. It takes a lot of courage. It takes patience, and understanding. It takes failure. That will be a life-long process for you, and for your children. From the first interaction you have with them to the most recent phone call (or emoji) you exchanged. Be attentive, look, listen, and learn.

All together now.

I have to say, now more than ever we need to build on each other's strengths. If you, like me, belong to this in-between generation, we are key players. We are able to connect and integrate the best of our elders and download it into the newer generations. We can still translate the lessons from the past and enable young people to connect to their humanity. All of our life experiences as a collective will make us better and stronger.

As I write this, I keep running into desperate stories of parents who did not see how their children got hooked into toxic online behavior. Anywhere from sexting to joining groups that practice bulimic activities and bullying, parents do not know where they failed.

Let me tell you, you did not fail. We all did. We said it takes a village, and it does. We all need to do a better job. We all need to help and ask for help. We are all key to get children ready for a more digitalized era, while empowering their humanity. Just as we need to bring technology to those who are not yet ready or able to use it. Companies and developers need to give us better tools to monitor, and we need to monitor. Parents need to talk to their children, and children need to practice their best judgement and take care of each other.

In my social media I tend to use the phrase "bringing social back to social media", but the truth is we need to "bring human back". Let us make human a good word now. Too long we have made mistakes and bad choices because "we didn't know" better. Now we do, and if we do not, we can Google it! Be informed, and learn how to discern information. Then teach others how to do it. Pass knowledge along, and pass values along, we all need it.

Obviously, nothing works if we are not open to learning. So, just as much as you put out into the world, keep your eyes and ears open. Remember that our human experience is built from what we see around us, and from what we know within us, so mix your own experience with what you learn in other people's life lessons.

People first.

You can fool yourself all you want. You can tell yourself that you can multi-task and you have everything under control. You can swear and prove over and over again that you can drive and text, or drive and record a video, or maybe cook and check your Facebook. The truth is you cannot do more than one thing with your full and undivided attention.

We have been caught in between, and it seems like we took it to all areas of our life. I know I have done it. I know I have been watching TV and scrolling the timeline in my social media. I know I have been talking to someone in a meeting and replying to something in my phone. I have walked checking my phone and ran into walls or tripped with things on the street. I have been playing with my dogs and posting something online at the same time. I have picked up my phone even when it doesn't sound or notify me of something, just to see if there is something going on there, while I'm doing something else.

Let's be clear. We need to focus on one thing, and be clear about what is important at the time. Truth be told we usually do two things because we do not know how to choose which one is more important. If your children are important, and you get a message or a call, dismiss it. If your work is important and you get a message that needs to be answered, then step aside or go to another room, and take care of it.

You might be great at it right now, but time will pass. You will lose your ability to pay attention, because you are just not used to it anymore.

Do not lie to yourself. Just because something takes two seconds, it doesn't mean that you can ignore the person right in front of you. We seem to justify not being present because it is going to take just a minute. Acknowledge the people around you. Respect the presence and importance that others give to you, by giving it back.

Choose what needs to be done over what you want to do, and once you are done, fully focus on what comes next. And if it means you have to pay attention to your phone, then be kind to the person you are neglecting. If it is your children, put them in a safe situation. If it is your dog, at least put the leash on. If it is your family, excuse yourself and explain why you have to step out for a second or reply to that text.

The beauty of living caught in between both worlds is that we see the value of both. But just as with everything else, we have a choice. We want quick, simple, and effective. We want personal touch. We want deep relationships. We know what we have to do.

The platform may not make as much of a difference as our own approach to what we want to do.

Be present, and take the best out of both worlds. If you do so, you may be bringing the best of yourself to both worlds as well.

Always put people first, more so if they are people you love.

www.ingramcontent.com/pod-product-compliance
Lightning Source LLC
Chambersburg PA
CBHW030457220526
45464CB00006B/2565